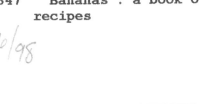

# BANANAS

# BANANAS

## *A Book of Recipes*

INTRODUCTION BY KATE WHITEMAN

LORENZ BOOKS

NEW YORK • LONDON • SYDNEY • BATH

EF SP CERS VWM SH VRW LK

First published by Lorenz Books in 1997

© 1997 Anness Publishing Limited

Lorenz Books is an imprint of
Anness Publishing Inc.
27 West 20th Street
New York, New York 10011

ISBN 1 85967 487 9

*Publisher* Joanna Lorenz
*Senior Cookery Editor* Linda Fraser
*Project Editor* Anne Hildyard
*Designer* Bill Mason
*Illustrations* Anna Koska

*Photographers* Karl Adamson, Steve Baxter, James Duncan, Amanda Heywood,
Don Last, Patrick McLeavey and Thomas Odulate
*Recipes* Kit Chan, Christine France, Sarah Gates, Shirley Gill, Rosamund Grant,
Manisha Kanani, Sallie Morris, Anne Sheasby and Steven Wheeler
*Food for photography* Carla Capalbo, Joanne Craig, Jane Hartshorn, Wendy Lee, Jane
Stevenson, Judy Williams and Elizabeth Wolf-Cohen.
*Stylists* Madeleine Brehaut, Hilary Guy, Blake Minton, Kirsty Rawlings
and Fiona Tillett
*Jacket photography* Janine Hosegood

Printed and bound in China


1 3 5 7 9 10 8 6 4 2

# Contents

# $\mathscr{I}$NTRODUCTION

Bananas are the ultimate healthy, high-energy convenience food. Neatly packaged in their attractive easy-to-open skins, the sweet-tasting fruit with their delicate aroma are packed with vitamins, minerals, and complexion-enhancing pectin, and are high in energy-giving starch. Bananas originated in South-East Asia and have grown in the tropics since ancient times. Hundreds of different varieties of bananas flourish in many tropical countries, from the Caribbean and Africa to South-East Asia, South America, and the Canaries. The very sight of huge "hands" of bananas pointing upward through the elongated green leaves of the banana plant is enough to conjure up a feeling of exoticism. All sweet bananas are delicious eaten raw, but their versatile flesh is equally good made into ice cream, soufflés, and trifles. They combine well with other ingredients from tropical regions, especially brown sugar, rum, coconut, mangoes, passion fruit, chocolate, and pineapple.

Bananas can also be used as a vegetable. You might mistake green-skinned plantains for unripe sweet bananas, but they are starchier and contain less sugar, so are almost always served cooked in savory dishes. They can be boiled, fried, or mashed like potatoes and are an essential ingredient of many West Indian and African dishes. They have a particular affinity with white fish and bacon. Every part of the banana can be used in cooking, even the skin, which forms a protective wrapping for baked bananas, while the leaves make neat aromatic packets to enclose a filling.

Because they come from the tropics, bananas are available all year round. They are exported while they are still green, and are ripened in storage. You can buy bananas in various stages of ripeness, from hard and green to soft and yellow mottled with brown, but they ripen very quickly, turning brown with speckled yellow flesh which is really only suitable for cooking.

This book begins with an introduction to the various bananas available in stores and supermarkets, and guides you through choosing and preparing them. Two chapters of exciting savory dishes follow, including exotic appetizers, snacks, and main courses from Africa, Thailand, and the Caribbean. Finally, dessert bananas make their appearance in many attractive guises, from healthy tropical banana fruit salad to irresistible chocolate cake with banana sauce, and including combinations like Brazilian coffee bananas, hot bananas with rum and raisins, and banana and passion fruit whip. All these recipes reveal the versatility of the world's favorite exotic fruit.

# Types of Bananas

## Dessert bananas

Standard dessert bananas are long, curved fruit with yellow skins, turning to speckled brown as they ripen. Unripe bananas have green skins or greenish tips. They are difficult to peel and are not pleasant to eat, as they have a crunchy texture and make your mouth pucker. The most common type of dessert banana is Cavendish, which has many sub-varieties. However, unless you are a banana expert, it is virtually impossible to differentiate between them. All dessert bananas are delicious raw or cooked.

## Apple bananas

These tiny yellow bananas from Colombia are not much bigger than a woman's finger. They have golden flesh with a faint aroma and flavor of apple.

## Green bananas

Large green bananas can be cooked in much the same way as potatoes and are often served as a substitute, although they have a blander flavor. They have crisp flesh, which can be boiled, mashed, or fried in butter. Fried green banana rounds are good served with white fish or in a curry.

## Banana leaves

Banana leaves are shaped like elongated fans, with long straight veins running from the center to the edges. Whole unblemished banana leaves are soft and malleable, and can be folded or rolled into packets to keep the filling moist and succulent. They impart a delicate flavor to the food and look very attractive.

## Plantains

Yellow and green plantains look like large bananas, but their shape is longer and flatter. They have firm pinkish flesh which contains more starch but is less sweet than that of dessert bananas. As the plantain ripens, the flesh becomes darker and sweeter and can be used in desserts. However, it is more usual to serve plantains as a vegetable. Very firm plantains can be thinly sliced and deep-fried as an unusual alternative to potato chips.

## Dried bananas

Drying intensifies the sweetness of bananas. Dried bananas are usually sold as bars; they are dark, sticky, and very sweet. Eaten on their own, they make a chewy but nutritious and energizing snack.

Unripe plantain

Green bananas

Dried bananas

Ripe plantain

Apple bananas

Dessert bananas

Banana leaf

# $\mathcal{B}$ASIC $\mathcal{T}$ECHNIQUES

## CHOOSING BANANAS

• Uniformly green fruit is unripe and almost inedible.

• Bananas with green-tinged ends are slightly unripe, with a crisp texture and a refreshing taste.

• Perfectly ripe bananas are uniformly yellow and the flesh is soft and sweet.

• As bananas continue to ripen, brown patches appear on the skin until it is covered with brown mottling. At this stage, the flesh is very sweet and soft, and perfect for mashing for mousses or banana sandwiches, for example.

• The final stage, when the skin is dark brown all over, is the last chance to eat the banana. The flesh will have almost collapsed and tastes like fermenting honey. A banana which has reached this stage of over-ripeness is best used for cooking.

## PREPARING BANANAS

PEELING BANANAS

*Strip the skin from ripe bananas in complete sections from the stalk end. Remove any white threads from the flesh before eating or using the bananas in cooking.*

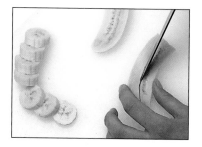

SLICING BANANAS

*For fruit salads and trifles, slice the bananas crosswise in rounds. If you are baking the bananas, slice them lengthwise, then cut in half again into chunks.*

DIPPING IN LEMON JUICE

*After peeling and slicing bananas, dip them into fresh lemon juice to prevent the flesh discoloring. Turn the banana slices in the lemon juice to coat them on all sides.*

## PEELING PLANTAINS

*Top and tail the plantains using a small, sharp knife. Discard the ends and stalks, and cut the plantains in half crosswise.*

*Using a small, sharp knife, slit the skin of each piece of plantain, in a few places, along the natural ridges. Take care not to cut through the flesh.*

*Ease up the edge of the skin and run the tip of your thumb along the plantain pieces, lifting the skin. Peel off the skin and discard it.*

## SESAME TOFFEE BANANAS

Serves 4

*The contrast between the crunchy coating and the soft banana within is irresistible.*

Cut four peeled bananas into 1in chunks and brush with lemon juice. Sift ½ cup flour and a pinch of salt into a bowl and add 1 egg and ½ cup milk. Stir to make a batter.

Put 1 tbsp peanut oil, 1⅛ cup superfine sugar, ¼ cup water, and 2 tbsp sesame seeds in a saucepan and heat gently until the sugar dissolves. Increase the heat and boil for 10–15 minutes to make a deep golden caramel.

Heat a pan of vegetable oil to 375°F for deep-frying. Dip the bananas in the batter and fry them in batches until puffed up. Remove and drain on paper towels. Using two forks, dip the banana fritters into the sesame caramel to coat them all over.

# Appetizers, Snacks, and Side Dishes

*Bananas and plantains add delicious flavor and texture to a variety of exotic dishes, such as crunchy assorted plantain appetizers, steamed fish packets, and green bananas and yam.*

# PLANTAIN AND CORN CHOWDER

*Unlike dessert bananas, plantains are always cooked. Firm and starchy, they make a delicious soup.*

**Serves 4**

2 tbsp butter or margarine

1 onion, finely chopped

1 garlic clove, crushed

10oz yellow plantains, peeled and
   sliced

1 large tomato, skinned and chopped

1 cup corn kernels

1 tsp crushed dried tarragon

3¾ cups vegetable or chicken stock

1 fresh green chili, seeded
   and chopped

pinch of grated nutmeg

salt and ground black pepper

Melt the butter or margarine in a saucepan over a medium heat, add the onion and garlic, and fry for 3–4 minutes until the onion is soft.

Add the sliced plantains, tomato, and corn and cook for 5 minutes.

Stir in the tarragon, vegetable or chicken stock, and chili, with salt and pepper to taste. Bring to a boil, then lower the heat, and simmer for about 10 minutes or until the plantains are just tender. Stir in the nutmeg and serve the soup at once.

# STEAMED BANANA LEAF PACKETS

*Very neat and delicate, these seafood packets from Thailand make an excellent appetizer or light lunch.*

**Serves 4**

*8oz crab meat*

*2oz peeled shrimp, chopped*

*6 drained water chestnuts, chopped*

*2 tbsp chopped bamboo shoots*

*1 tbsp chopped scallion*

*1 tsp chopped fresh ginger root*

*2 tbsp soy sauce*

*1 tbsp fish sauce*

*12 rice sheets*

*banana leaves, for lining steamer*

*oil, for brushing*

*2 scallions, shredded, 2 fresh red chilies, seeded and sliced, and cilantro leaves, to garnish*

### COOK'S TIP

*The seafood packets will spread out when cooked so be sure to space them well apart in the steamer to prevent them sticking together.*

Combine the crab meat, chopped shrimp, chestnuts, bamboo shoots, scallion, and ginger in a bowl. Mix well, then add 1 tbsp of the soy sauce and all the fish sauce. Stir until blended.

Take a rice sheet and dip it in warm water. Place it on a flat surface and leave for a few seconds to soften.

Place a spoonful of the filling in the center of the sheet and fold into a square packet. Repeat with the rest of the rice sheets and seafood mixture.

Use banana leaves to line a steamer, then brush them with oil. Place the packets, seam side down, on the leaves and steam over a high heat for 6–8 minutes or until the filling is cooked. Transfer to a plate and garnish with the scallions, chilies, and cilantro leaves.

# ASSIETTE OF PLANTAINS

*This mélange of succulent sweet and savory plantains makes a delicious crunchy appetizer.*

**Serves 4**

*vegetable oil, for shallow frying*

*2 green plantains*

*½ onion*

*1 yellow plantain*

*pinch of garlic granules*

*salt and cayenne pepper*

Heat the oil in a large skillet over a medium heat. While the oil is heating, peel one of the green plantains and cut into very thin rounds using a vegetable peeler. Fry the plantain rounds in the oil for about 3 minutes, turning until golden brown. Drain on paper towels and keep warm.

Coarsely grate the remaining green plantain onto a plate. Slice the onion into wafer-thin shreds and mix with the grated plantain. Heat a little more oil in the skillet and fry handfuls of the mixture for 2–3 minutes, until golden, turning once. Drain on paper towels and keep warm.

Peel the yellow plantain, cut it in half lengthwise, and dice. Sprinkle with the garlic granules and cayenne pepper. Heat a little more oil in the skillet and fry the plantain until evenly golden brown. Drain on paper towels and then arrange the three varieties of cooked plantains in shallow dishes. Sprinkle with salt and serve as a snack.

# FRIED YELLOW PLANTAINS

*When plantains are ripe, they turn yellow and become sweeter, but not as sweet as dessert bananas. This makes them the perfect accompaniment to broils and roasts.*

**Serves 4**
*2 yellow plantains*
*oil, for shallow frying*
*finely snipped fresh chives or finely*
*  chopped fresh mixed herbs,*
*  to garnish*

Using a small sharp knife, top and tail the plantains, and cut each of them in half. Slit the skin only, along the natural ridges of each piece of plantain. Ease up the edge of the skin and run the tip of your thumb along the plantains, lifting the skin. Peel away the skin and slice the plantains in half lengthwise.

Heat a little oil in a large skillet and fry the plantain slices for 2–3 minutes on each side until golden brown. Do not overcook as they can become rather dry and starchy.

When the plantains are golden and crisp, lift out the slices using a perforated spoon. Drain the plantains on paper towels to remove excess oil.

Serve hot or cold, sprinkled with finely snipped chives or with finely chopped fresh mixed herbs. Serve the fried plantains with broiled or roast meat, fish, or vegetarian dishes.

### COOK'S TIP

*Look for plantains at vegetable markets and large supermarkets. You can use bananas as a substitute, but they tend to become rather too soft when cooked.*

# GREEN BANANAS AND YAM

*For a tantalizing taste of the tropics, try this unusual vegetable and banana medley.*

**Serves 3–4**

4 green bananas, peeled and halved
1lb white yam, peeled and cut into
  pieces
1 thyme sprig
1½oz creamed coconut, cubed
salt and ground black pepper
chopped fresh thyme and sprigs of
  thyme, to garnish

### COOK'S TIP
*Creamed coconut is available
in blocks from oriental food
stores and some supermarkets.
It can be thinned by mixing it
with a little water to make
coconut milk.*

Bring 3¾ cups water to a boil in a large saucepan, lower the heat, and add the green bananas and yam. Simmer gently for about 10 minutes.

Add the thyme and coconut, with salt and pepper to taste. Bring back to a boil and cook over a medium heat until the yam and banana are tender.

Using a perforated spoon, transfer the yam and banana to a plate. Continue cooking the coconut milk, stirring frequently, until it is thick and creamy in consistency.

When the sauce is ready, return the vegetables to the pan and heat through. Spoon into a warm serving dish, sprinkle with chopped thyme, and garnish with sprigs of thyme.

# YAM AND PLANTAIN FU FU

*Small, savory, and packed with flavor, serve these yam and plantain balls with casseroles and stews.*

**Serves 4**

*1lb white yam*

*2 green plantains*

*1 tbsp butter or margarine*

*salt and ground black or white*
*   pepper*

*flat leaf parsley, to garnish*

Peel, wash, and slice the yam. Place in a saucepan and pour in lightly salted cold water to cover. Cut the green plantains in half, slit along the natural ridges in three places, and remove the skins. Add to the yam, bring to a boil, and cook for 25 minutes until the vegetables are tender.

Drain the vegetables and place in a blender or food processor. Add the butter or margarine, season well with salt and pepper, and process until smooth and lump free.

Scrape the fu fu into a bowl, then take small handfuls, and shape into balls. Reheat in a low oven or in the microwave. Garnish with parsley and serve.

# Main Courses

*Bananas and plantains combine surprisingly
well with fish and vegetables in spicy curries,
colorful kebabs, and succulent salads. Banana
leaves, wrapped around fish, add their own
distinctive flavor and aroma.*

# BANANA CURRY

*The sweetness of the bananas combines well with the spices used to produce a mild, sweet curry.*

**Serves 4**

*4 under-ripe bananas*

*2 tbsp ground coriander*

*1 tbsp ground cumin*

*1 tsp chili powder*

*½ tsp salt*

*¼ tsp ground turmeric*

*1 tsp sugar*

*1 tbsp gram flour*

*3 tbsp chopped fresh cilantro*

*6 tbsp corn oil*

*¼ tsp cumin seeds*

*¼ tsp black mustard seeds*

*fresh sprigs of cilantro, to garnish*

*chappatis, to serve*

### COOK'S TIP

*Choose bananas that are slightly under-ripe so that they retain their shape and do not become unpleasantly mushy when they are cooked.*

Trim the bananas leaving the skin on, and cut each into three equal pieces. Make a lengthwise slit in each piece of banana, without cutting through.

Mix the coriander, cumin, chili powder, salt, turmeric, sugar, gram flour, and chopped cilantro in a soup plate. Stir in 1 tbsp of the oil. Carefully stuff each piece of banana with the spice mixture, taking care not to break them in half.

Heat the remaining oil in a large heavy-based saucepan and fry the cumin and mustard seeds for 2 minutes or until they begin to splutter. Add the bananas and toss gently in the oil. Cover and simmer over a low heat for 15 minutes, stirring from time to time, until the bananas are soft, but not mushy. Garnish with the fresh cilantro and serve with warm chappatis.

# PLANTAIN AND VEGETABLE KEBABS

*Tasty and colorful, these kebabs make a delightful main course for vegetarians or can be served as a side dish.*

**Serves 4**

*4oz pumpkin, peeled and cubed*

*1 red onion, cut into wedges*

*1 small zucchini, sliced*

*1 yellow plantain, sliced*

*1 eggplant, diced*

*½ red bell pepper, seeded and diced*

*½ green bell pepper, seeded and diced*

*12 button mushrooms, trimmed*

*4 tbsp lemon juice*

*4 tbsp olive or sunflower oil*

*3–4 tbsp soy sauce*

*⅔ cup tomato juice*

*1 fresh green chili, seeded*
  *and chopped*

*½ onion, grated*

*3 garlic cloves, crushed*

*1½ tsp dried tarragon,*
  *crushed*

*¾ tsp each dried basil, dried thyme,*
  *and ground cinnamon*

*2 tbsp butter*

*1¼ cups vegetable stock*

*freshly ground black pepper*

Place the pumpkin in a small bowl and cover with boiling water. Blanch for 2–3 minutes, then drain, refresh under cold water, drain again, and tip into a large bowl. Add the red onion, zucchini, plantain, eggplant, bell peppers, and mushrooms.

Mix the lemon juice, oil, soy sauce, tomato juice, chili, grated onion, garlic, herbs, cinnamon, and black pepper in a pitcher. Pour over the vegetables. Toss well, then set aside in a cool place to marinate for 3–4 hours.

Drain the vegetables and thread them alternately onto eight skewers. Broil under a low heat for about 15 minutes, turning the kebabs frequently, until golden brown. Baste occasionally with the marinade to keep the vegetables moist.

Place the remaining marinade, butter, and stock in a pan and bring to a boil. Lower the heat and simmer for 10 minutes to cook the onion and reduce the sauce. Pour into a serving pitcher. Arrange the vegetable skewers on a plate. Serve with a rice dish or salad.

# PLANTAIN AND GREEN BANANA SALAD

*Cooking plantains and bananas in their skins helps to retain the soft texture so that they absorb all the flavor of the dressing.*

**Serves 4**

*2 ripe yellow plantains*

*3 green bananas*

*1 garlic clove, crushed*

*1 red onion*

*1–2 tbsp chopped fresh*
  *cilantro*

*3 tbsp sunflower oil*

*1½ tbsp malt vinegar*

*salt and coarse-grain black pepper*

Slit the plantains and bananas lengthwise along their natural ridges, then cut in half, and place in a large saucepan. Pour in water to cover, add a little salt, and bring to a boil.

Boil the plantains and bananas gently for 20 minutes until tender, then drain well. When they are cool enough to handle, peel and cut them into medium-size slices.

Put the plantain and banana slices into a bowl and add the crushed garlic, turning to mix.

Cut the onion in half and slice it thinly. Add to the bowl with the chopped fresh cilantro, oil, and vinegar. Add salt and pepper to taste. Toss to mix, then serve.

**COOK'S TIP**

*Red onions are mild and sweet, so they are ideal for mixing in salads and for adding extra flavor to sandwiches.*

# SPINACH PLANTAIN ROUNDS

*This delectable way of serving plantains is a little effortful to make, but well worth the trouble. The plantains must be ripe, but still firm.*

**Serves 4**

*2 large yellow plantains, peeled*
*oil, for frying*
*2 tbsp butter*
*1 tbsp finely chopped onion*
*2 garlic cloves, crushed*
*1lb fresh spinach, chopped*
*pinch of freshly grated nutmeg*
*1 egg, beaten*
*whole wheat flour, for dipping*
*salt and ground black pepper*

Using a small sharp knife, carefully cut each plantain lengthwise into four slices. Heat a little oil in a large skillet and fry the slices on both sides until pale gold in color, but not fully cooked. Lift out and drain on paper towels, and reserve the oil in the skillet.

Melt the butter in a saucepan and sauté the onion and garlic for 2–3 minutes until the onion is soft. Add the spinach and nutmeg, with salt and pepper to taste. Cover and cook for about 5 minutes until the spinach has reduced. Cool, then tip into a strainer, and press out any excess moisture.

Curl the plantain slices into rings and secure each ring with a wooden toothpick. Pack each ring with a little of the spinach mixture.

Place the egg and flour in two separate dishes. Add a little more oil to the skillet, if necessary, and heat until moderately hot. Dip the spinach and plantain rounds in the egg, and then in the flour. Fry on both sides for 1–2 minutes until golden brown. Drain on paper towels and serve hot or cold.

**COOK'S TIP**

*If fresh spinach is not available, use frozen spinach. Thaw completely and drain thoroughly in a strainer before cooking.*

# SPICY PLANTAINS WITH YAM

*This tomato-flavored plantain dish is a perfect partner for a spicy meat or fish stew.*

**Serves 4**

*2 green plantains*

*1lb white yam*

*2 tomatoes, skinned and chopped*

*1 fresh red chili, seeded and chopped*

*1 onion, chopped*

*½ vegetable stock cube*

*1 tbsp palm oil*

*1 tbsp tomato paste*

*salt*

### VARIATION

*Use 2 leeks or shallots instead of the onion, and substitute diced summer squash or zucchini for the yam.*

Peel the plantains and cut into six rounds, then peel and dice the yam. Place in a large saucepan. Add 2½ cups water, bring to a boil, and cook for 5 minutes.

Add the tomatoes, chili, and onion to the pan and simmer for about 10 minutes, then crumble in the stock cube, stir well, cover, and simmer for another 5 minutes.

Stir in the oil and tomato paste and continue cooking for about 5 minutes until the plantains are tender. Season with salt and pour into a warm serving dish. Serve immediately.

# BAKED FISH IN BANANA LEAVES

*Fish baked in banana leaves is particularly succulent and flavorful. This is a great dish for barbecuing.*

**Serves 4**

*1 cup coconut milk*

*2 tbsp red curry paste*

*3 tbsp fish sauce*

*2 tbsp superfine sugar*

*5 kaffir lime leaves, torn*

*4 fish fillets, about 6oz each*

*6oz mixed vegetables, such as carrots
    or leeks, finely shredded*

*4 banana leaves*

**For the garnish**

*2 tbsp shredded scallions*

*2 fresh red chilies, finely sliced*

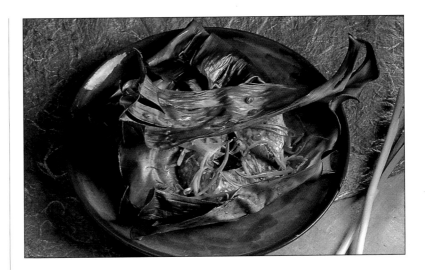

Combine the coconut milk, curry paste, fish sauce, sugar, and kaffir lime leaves in a shallow dish. Add the fish and marinate for 15–30 minutes. Preheat the oven to 400°F.

Mix the selected vegetables together and place a fourth of the mixture on top of a banana leaf. Place a fish fillet on top of each and moisten it with a little of its marinade.

Wrap the fish up by turning in the sides and ends of the leaf and securing the package with toothpicks. Repeat with the rest of the leaves, vegetables, and fish.

Bake for 20–25 minutes or until the fish is cooked. Alternatively, cook under the broiler or in a hinged broiler over a barbecue. Just before serving, garnish the fish with a sprinkling of scallions and sliced red chilies.

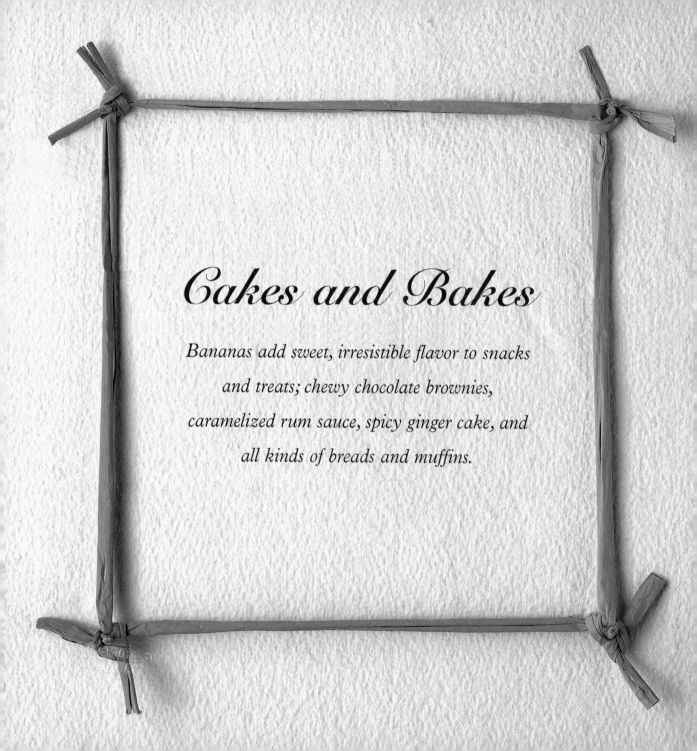

# Cakes and Bakes

*Bananas add sweet, irresistible flavor to snacks
and treats; chewy chocolate brownies,
caramelized rum sauce, spicy ginger cake, and
all kinds of breads and muffins.*

# CHOCOLATE AND BANANA BROWNIES

*Bananas give brownies a delicious flavor and keep them marvelously moist.*

**Makes 9**

5 tbsp cocoa powder

1 tbsp superfine sugar

5 tbsp milk

3 large bananas, mashed

1 cup brown sugar

1 tsp vanilla extract

5 egg whites

¾ cup self-rising flour

⅔ cup oat bran

1 tbsp confectioner's sugar, for
   dusting

**COOK'S TIP**

*Store these brownies in an airtight container for one day before eating them – their flavor becomes stronger and improves with keeping.*

Preheat the oven to 350°F. Line an 8in square baking pan with baking parchment. In a bowl, mix the cocoa powder and superfine sugar with the milk. Add the bananas, brown sugar, and vanilla extract. Mix well.

In a mixing bowl, beat the egg whites lightly with a fork. Add the chocolate mixture and continue to beat well. Sift the flour over the mixture and fold in with the oat bran. Pour into the prepared pan.

Bake for 40 minutes or until firm. Cool in the pan for 10 minutes, then turn out onto a wire rack, and cool completely. Cut into nine wedges and dust lightly with confectioner's sugar before serving.

# CHOCOLATE CAKE WITH BANANA SAUCE

*Caramelized banana and rum sauce tastes superb with wedges of chocolate cake.*

**Serves 6**

*4oz dark chocolate, broken into squares*

*½ cup sweet butter, at room temperature*

*1 tbsp instant coffee powder*

*5 eggs, separated*

*1 cup sugar*

*1 cup all-purpose flour*

*1 tsp ground cinnamon*

**For the sauce**

*4 ripe bananas*

*4 tbsp brown sugar*

*1 tbsp lemon juice*

*¾ cup heavy cream*

*1 tbsp rum (optional)*

Preheat the oven to 350°F. Grease an 8in round cake pan. Bring a small saucepan of water to a boil. Remove it from the heat and place a heatproof bowl on top. Add the chocolate and butter to the bowl and leave until melted, stirring occasionally. Stir in the coffee powder and set aside.

Mix the egg yolks and sugar in a bowl. Beat by hand or with an electric mixer until thick and lemon-colored. Add the chocolate mixture and beat on low speed for just long enough to blend the mixtures evenly.

Sift the flour and cinnamon into a bowl. In another bowl, whisk the egg whites to stiff peaks. Fold a spoon of egg white into the chocolate mixture to lighten it. Fold in the remaining egg white in batches, alternating with the sifted flour mixture.

Pour the mixture into the prepared pan. Bake for 40–50 minutes or until a skewer inserted in the center comes out clean. Turn out onto a wire rack.

Preheat the broiler. Make the sauce. Slice the bananas into a shallow, flameproof dish. Add the brown sugar and lemon juice and stir to mix. Place under the broiler and cook, stirring occasionally, for about 8 minutes until the sugar is caramelized and bubbling. Mash the bananas into the sauce until almost smooth. Stir in the cream and rum, if using. Slice the cake and serve it warm, with the sauce.

# BANANA MUFFINS

*Make plenty of these delectable treats – banana muffins are irresistible at any time of the day.*

**Makes 10**

*2 cups all-purpose flour*

*1 tsp baking powder*

*1 tsp baking soda*

*¼ tsp salt*

*¼ tsp grated nutmeg*

*½ tsp ground cinnamon*

*3 large ripe bananas*

*1 egg*

*⅓ cup brown sugar*

*4 tbsp vegetable oil*

*¼ cup raisins*

Preheat the oven to 375°F. Line 10 muffin cups with paper liners or grease them lightly. Sift the flour, baking powder, baking soda, salt, nutmeg, and cinnamon into a bowl. Set aside.

Mash the bananas in a mixing bowl until creamy. Using a hand-held electric mixer, beat in the egg, sugar, and oil. Add the dry ingredients and mix until just blended. Stir in the raisins.

Fill the muffin cups two-thirds full. Bake for 20–25 minutes or until the tops spring back when lightly touched. Transfer the muffins to a wire rack to cool slightly. Serve warm.

**COOK'S TIP**

*If there are any empty cups in the muffin tray when you have used up the mixture, fill them with water before placing the tray in the oven to ensure that the muffins bake evenly.*

# BANANA GINGER CAKE

*Bananas and ginger make a winning combination. This cake actually improves with keeping.*

**Makes 12 bars**

1¾ cups all-purpose flour

2 tsp baking soda

2 tsp ground ginger

1¾ cups medium oatmeal

4 tbsp dark brown sugar

6 tbsp butter or margarine

⅔ cup corn syrup

1 egg, beaten

3 ripe bananas, mashed

¾ cup confectioner's sugar

preserved ginger, to decorate

### COOK'S TIP
*This is a nutritious, energy-giving cake that is a good choice for brown-bag lunches, as it doesn't break up or crumble very easily.*

Preheat the oven to 325°F. Grease and line an 11 x 7in cake pan with baking parchment. Sift the flour, baking soda, and ground ginger into a bowl, then stir in the oatmeal.

Melt the sugar, butter or margarine, and syrup in a saucepan, then stir into the flour mixture. Beat in the egg and mashed bananas.

Spoon the mixture into the pan and bake for about 1 hour, or until firm to the touch. Allow to cool in the pan, then turn out, and cut into bars.

Sift the confectioner's sugar into a bowl and stir in just enough water to make a smooth, runny frosting. Drizzle the frosting over each square and top with slices of preserved ginger.

# BANANA AND LEMON CAKE

*Light, moist, and flavorsome, this cake keeps very well and is everybody's favorite.*

**Serves 8–10**

2¼ cups all-purpose flour

1¼ tsp baking powder

pinch of salt

½ cup sweet butter, at room
    temperature

scant 1 cup superfine sugar

½ cup brown sugar

2 eggs

½ tsp grated lemon rind

1 cup mashed, very
    ripe bananas

1 tsp vanilla extract

4 tbsp milk

¾ cup chopped walnuts

lemon-rind curls, to decorate

**For the frosting**

½ cup butter, at room
    temperature

4½ cups confectioner's sugar

1 tsp grated lemon rind

3–5 tbsp lemon juice

Preheat the oven to 350°F. Grease two 9in round cake pans, and line the bases with baking parchment. Sift the flour, baking powder, and salt into a bowl.

Beat the butter and sugars in a large mixing bowl, until light and fluffy. Beat in the eggs, one at a time, then stir in the grated lemon rind.

Mix the mashed bananas with the vanilla extract and milk in a small bowl. Stir this, in batches, into the creamed butter mixture, alternating with the sifted flour. Stir lightly until just blended. Fold in the walnuts.

Divide the mixture between the cake pans and spread evenly. Bake for 30–35 minutes, until a skewer inserted in the center comes out clean. Leave to stand for 5 minutes before turning out onto a wire rack. Peel off the lining parchment and leave to cool.

Make the frosting. Cream the butter in a bowl until smooth, then gradually beat in the confectioner's sugar. Stir in the lemon rind and enough of the lemon juice to make a spreading consistency.

Place one of the cakes on a serving plate. Spread over one-third of the frosting, then top with the second cake. Spread the remaining frosting evenly over the top and sides of the cake. Decorate with lemon-rind curls.

# BANANA NUT BREAD

*Banana bread is always popular. This delicious, healthy version has added pecans.*

**Makes 1 loaf**

½ cup sweet butter, at room
   temperature

½ cup sugar

2 eggs, at room temperature

1 cup all-purpose flour

1 tsp baking soda

¼ tsp salt

1 tsp ground cinnamon

½ cup whole wheat flour

3 large ripe bananas

1 tsp vanilla extract

½ cup pecan nuts, chopped

**COOK'S TIP**

*If the cake mixture shows signs
of curdling when you add the
eggs, beat in a little of the sifted
flour mixture.*

Preheat the oven to 350°F. Line the bottom and sides of a 9 x 5in loaf pan with baking parchment.

Using an electric mixer, cream the butter and sugar in a bowl until light and fluffy. Add the eggs, one at a time, beating well after each addition.

Sift the all-purpose flour, baking soda, salt, and cinnamon over the butter mixture. Stir in thoroughly, then stir in the whole wheat flour.

Mash the bananas to a purée, then stir into the mixture. Stir in the vanilla extract and pecans. Pour into the prepared pan and level the surface.

Bake the loaf for 50–60 minutes, until a skewer inserted in the center comes out clean. Turn out onto a wire rack to cool.

# Glazed Banana Spice Loaf

*Bananas and spices are perfect partners in this delicious loaf, which makes an ideal family treat.*

**Makes 1 loaf**

*1 large ripe banana*

*½ cup butter, at room temperature*

*scant ¾ cup superfine sugar*

*2 eggs, at room temperature*

*1¼ cups all-purpose flour*

*1 tsp salt*

*1 tsp baking soda*

*½ tsp grated nutmeg*

*¼ tsp ground allspice*

*¼ tsp ground cloves*

*¾ cup sour cream*

*1 tsp vanilla extract*

**For the glaze**

*1⅛ cups confectioner's sugar*

*1–2 tbsp lemon juice*

Preheat the oven to 350°F. Line an 8½ x 4½in loaf pan with wax paper and lightly grease. Using a fork, mash the banana in a bowl. Set aside. With an electric mixer, cream the butter and sugar until light and fluffy. Add the eggs, one at a time, beating well after each addition.

Sift together the flour, salt, baking soda, nutmeg, allspice, and cloves. Add to the butter mixture and stir to mix well. Add the sour cream, banana, and vanilla extract and mix just enough to blend. Pour into the prepared pan. Bake for 45–50 minutes, until the top springs back when lightly touched. Cool in the pan for 10 minutes, then leave on a wire rack to cool.

For the glaze, mix the confectioner's sugar and lemon juice, stirring until smooth. Place the cooled loaf on a rack set over a cookie sheet. Pour the glaze over the top of the loaf and allow to set.

# Hot Desserts

Cooked bananas have a sweetness and melting texture in delicious dishes such as crisp deep-fried bananas, hot bananas spiced with rum and cinnamon, and pancakes with lime and maple syrup or with chocolate chips and toasted almonds.

# BANANA MANDAZIS

*These delicious banana fritters come from Africa, where they are very popular.*

**Serves 4**

*1 egg*

*2 ripe bananas, roughly chopped*

*⅔ cup milk*

*½ tsp vanilla extract*

*2 cups self-rising flour*

*1 tsp baking powder*

*3 tbsp sugar*

*vegetable oil, for deep-frying*

*confectioner's sugar, for dusting*

### COOK'S TIP

*Drain each batch of mandazis well on paper towels and keep them hot in a low oven while you are cooking the remainder.*

Place the egg, bananas, milk, vanilla extract, flour, baking powder, and sugar in a blender or food processor. Process to a smooth, creamy batter. If it is too thick, add a little extra milk. Set aside for 10 minutes.

Heat the oil in a heavy-based saucepan or deep-fat fryer. When it is hot, place spoonfuls of the mixture in the oil and fry for 3–4 minutes until golden. Remove with a perforated spoon and drain. Keep hot while cooking the remaining mandazis. Dust with confectioner's sugar and serve at once.

# HOT BANANAS WITH RUM AND RAISINS

*Choose almost-ripe bananas with evenly colored skins, either all yellow or just green at the tips. Black patches indicate that the fruit is over-ripe.*

**Serves 4**

¼ *cup seedless raisins*

5 *tbsp dark rum*

4 *tbsp sweet butter*

4 *tbsp brown sugar*

4 *ripe bananas, peeled and halved lengthwise*

¼ *tsp grated nutmeg*

¼ *tsp ground cinnamon*

2 *tbsp slivered almonds, toasted*

*chilled cream or vanilla ice cream, to serve (optional)*

Put the raisins in a bowl and pour over the rum. Leave to soak for about 30 minutes, by which time the raisins will have plumped up.

Melt the butter in a skillet, add the brown sugar, and stir until just dissolved. Add the bananas and cook them for 4–5 minutes until they are just tender.

Sprinkle the nutmeg and cinnamon over the bananas, then pour over the rum and raisins. Stand back and carefully set the rum alight, using a long taper, and stir gently to mix.

Scatter the slivered almonds over the bananas and serve immediately with chilled cream or vanilla ice cream, if you like. Crème fraîche or Greek-style yogurt would also make a delicious accompaniment for the bananas.

> **VARIATION**
> *Try using golden raisins soaked in a tangerine-flavored liqueur, such as Van der Hum, or an orange-flavored liqueur, such as Grand Marnier, instead of raisins in rum.*

# SPICED NUTTY BANANAS

*Baked bananas are delectable however you serve them, but with a triple nut topping they are delicious.*

**Serves 3**

*6 ripe, but firm, bananas*

*2 tbsp chopped unsalted
 cashew nuts*

*2 tbsp chopped unsalted
 peanuts*

*2 tbsp shredded coconut*

*1 tbsp brown sugar*

*1 tsp ground cinnamon*

*½ tsp freshly grated nutmeg*

*⅔ cup orange juice*

*4 tbsp rum*

*1 tbsp butter or margarine*

*heavy cream or Greek-style yogurt,
 to serve*

### COOK'S TIP

*Freshly grated nutmeg makes
all the difference to this dish.
More rum can be added if
preferred. Chopped mixed nuts
can be used instead of peanuts.*

Preheat the oven to 400°F. Slice the bananas and place in a large, greased, shallow ovenproof dish. Do not leave for long at this stage as the bananas will discolor.

Mix the cashew nuts, peanuts, coconut, sugar, cinnamon, and nutmeg in a small bowl. Pour the orange juice and rum over the bananas, then sprinkle evenly with the nut and sugar mixture.

Dot the top evenly with butter or margarine. Bake for 15–20 minutes or until the bananas are golden brown and the sauce is bubbling.

Serve the bananas hot, with heavy cream or Greek-style yogurt.

# BANANAS FOSTER

*Possibly the most famous banana dessert, this originated in the French Quarter of New Orleans.*

**Serves 4**

¹⁄₃ *cup brown sugar*

¹⁄₂ *tsp ground cinnamon*

¹⁄₂ *tsp grated nutmeg*

*4 tbsp sweet butter*

*4 tbsp banana liqueur*

*5 tbsp dark rum*

*4 firm bananas*

*4 scoops firmly frozen vanilla*
   *ice cream*

### VARIATION
*You can ring the changes with praline, walnut, or even rum-and-raisin ice cream.*

Mix the sugar, cinnamon, and nutmeg in a bowl. Melt the butter in a heavy-based skillet and add the sugar and spice mixture, with the liqueur and rum. Stir over the heat until the sauce is syrupy.

Peel the bananas, cut them in half lengthwise, and add them to the skillet. Heat through, turning to coat with the sauce.

Tilt the skillet if you are cooking over gas to set light to the sauce. If your hob is electric, light the sauce with a long taper. Hold the skillet at arm's length while you do this.

As soon as the flames die down, put some pieces of banana on each plate, with a scoop of ice cream. Pour on the sauce and serve immediately.

# DEEP-FRIED BANANAS

*An Indonesian speciality, deep-fried bananas make a splendid spur-of-the-moment dessert.*

**Serves 8**

*1 cup self-rising flour*

*¼ cup rice flour*

*½ tsp salt*

*scant 1 cup water*

*finely grated lime rind (optional)*

*oil, for deep-frying*

*8 small bananas*

*superfine sugar, for dredging*

*1 lime, cut in wedges, to serve*

Sift both the flours and the salt together into a bowl. Add just enough water to make a smooth, coating batter. Mix well, then add the lime rind, if using.

Heat the oil to 375°F or until a cube of day-old bread browns in 30 seconds. Peel the bananas and dip them into the batter two or three times. Deep-fry until crisp and golden.

Drain the bananas and transfer them to a plate. Dredge with superfine sugar and serve hot, with the lime wedges.

**COOK'S TIP**

*Cook these delicious deep-fried bananas at the last minute before serving, so the crust is still crisp while the center stays melt-in-the-mouth soft.*

# BANANA, MAPLE, AND LIME PANCAKES

*Pancakes are a treat any day of the week, especially when they are filled with bananas and maple syrup.*

**Serves 4**

*1 cup all-purpose flour*

*1 egg white*

*1 cup milk*

*sunflower oil, for frying*

*strips of lime rind, to decorate*

**For the filling**

*4 bananas, sliced*

*3 tbsp maple syrup or*
*   corn syrup*

*2 tbsp fresh lime juice*

**M**ix the flour, egg white, and milk in a bowl. Add 4 tbsp cold water and beat until smooth and bubbly. Chill until needed.

Heat a little oil in a non-stick skillet and swirl in enough batter just to coat the base. Cook until golden, then turn over, and cook the other side. Keep hot while making the remaining pancakes.

Make the filling. Place the bananas, syrup, and lime juice in a pan and simmer gently for 1 minute. Spoon into the pancakes and fold into fourths. Sprinkle with strips of lime rind to decorate.

### COOK'S TIP

*Pancakes freeze well. To store for later use, stack and interleave them with baking parchment, overwrap with foil, and freeze for up to 3 months. Thaw thoroughly and reheat before using.*

# CHOCOLATE CHIP BANANA PANCAKES

*Serve these delicious banana pancakes as a dessert topped with cream and toasted almonds.*

**Makes 16**

*2 ripe bananas*

*scant 1 cup milk*

*2 eggs*

*1¼ cups self-rising flour*

*⅓ cup ground almonds*

*1 tbsp superfine sugar*

*pinch of salt*

*3 tbsp dark chocolate chips*

*butter, for frying*

**For the topping**

*⅔ cup heavy cream*

*1 tbsp confectioner's sugar*

*½ cup toasted slivered almonds*

### VARIATION

*For banana and blueberry pancakes, replace the chocolate with 1 cup fresh blueberries. Hot pancakes are simply delicious when they are served with ice cream.*

In a bowl, mash the bananas with a fork. Mix in half of the milk, then beat in the eggs. Sift in the flour and add the ground almonds, sugar, and salt. Mix lightly. Add the remaining milk and the chocolate chips to produce a thick batter.

Heat a knob of butter in a large non-stick skillet. Spoon the pancake mixture into heaps, allowing room for them to spread. When bubbles appear on top of the pancakes, turn them over, and cook briefly on the other side. Remove and keep hot.

Whip the cream lightly with the confectioner's sugar. Spoon onto the pancakes and top each with a few slivered almonds.

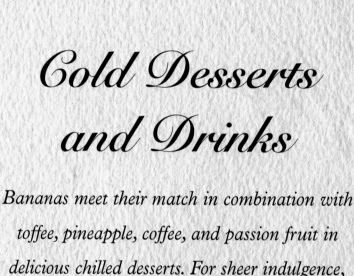

# Cold Desserts and Drinks

Bananas meet their match in combination with toffee, pineapple, coffee, and passion fruit in delicious chilled desserts. For sheer indulgence, there is nothing to beat rich banoffee pie washed down with a refreshing tropical fruit punch.

# BANANA AND MELON IN ORANGE VANILLA SAUCE

*A chilled banana and melon compote in a delicious orange sauce makes a perfect summer dessert.*

**Serves 4**

1¼ cups orange juice

1 vanilla bean

1 tsp finely grated orange rind

1 tbsp sugar

4 ripe, but firm, bananas

1 honeydew melon

2 tbsp lemon juice

strips of blanched orange rind, to
    decorate (optional)

### COOK'S TIP

*Most large supermarkets and
health food stores sell vanilla
beans. If they are unavailable,
use a few drops of vanilla
extract instead. Wash, dry, and
store the bean to use again.*

Place the orange juice in a small saucepan with the vanilla bean, orange rind, and sugar. Heat gently, stirring until the sugar has dissolved, then bring to a boil.

Lower the heat and simmer gently for 15 minutes or until the sauce is syrupy. Remove from the heat and leave to cool. Remove the vanilla bean. If using vanilla extract, stir into the sauce once it has cooled.

Roughly chop the bananas and melon, place in a large serving bowl, and toss with the lemon juice. Pour the cooled sauce over and chill the compote. Decorate with the blanched orange rind, if using, before serving.

# BRAZILIAN COFFEE BANANAS

*Rich, lavish, and sinful-looking, this banana dessert takes only moments to make!*

**Serves 4**

*4 small ripe bananas*

*1 tbsp instant coffee granules
 or powder*

*2 tbsp dark brown sugar*

*generous 1 cup Greek-
 style yogurt*

*1 tbsp toasted slivered almonds*

Peel and slice one banana. Peel and mash the remaining three in a bowl with a fork. Dissolve the coffee in 1 tbsp boiling water and stir into the mashed bananas.

Spoon a little of the mashed banana mixture into four serving dishes and sprinkle with sugar. Top with a spoonful of yogurt, then repeat the layers until all the ingredients are used up.

Using a skewer or toothpick, swirl the last layer of yogurt for a marbled effect. Finish with a few banana slices and slivered almonds. Serve cold, preferably within an hour of making.

### VARIATION

*For a special occasion, add a dash of dark rum or brandy, or crème de cacao to the bananas for extra richness.*

# QUICK BANANA PUDDING

*For instant energy and excellent taste, enjoy this simple banana pudding with a caramel topping.*

**Serves 6–8**

4 thick slices of ginger cake

6 ripe bananas

2 tbsp lemon juice

1¼ cups heavy cream

4 tbsp orange juice

2–3 tbsp brown sugar

### VARIATION

*Use fromage frais instead of heavy cream if you prefer. However, do not try to whip it – just stir in half the recommended amount of fruit juice.*

Break up the cake into chunks and arrange in an ovenproof dish. Slice the bananas into a bowl and toss with the lemon juice.

Whip the cream in a separate bowl until firm, then gently beat in the juice. Fold in the bananas and spoon the mixture over the ginger cake.

Top with the sugar, sprinkling it in an even layer. Place under a hot broiler for 2–3 minutes to caramelize. Chill in the refrigerator until set firm again if you wish, or serve at once.

# FLUFFY BANANA AND PINEAPPLE MOUSSE

*This light, low-fat banana mousse looks very impressive but is very easy to make.*

### Serves 6

*2 ripe bananas*

*1 cup cottage cheese*

*15oz can pineapple chunks or pieces*
   *in juice*

*1 sachet powdered gelatin*

*2 egg whites*

### COOK'S TIP

*For a simpler way of serving, use a 4-cup serving dish, which is able to hold all the mixture, and do not tie a collar around the top edge. Decorate the top of the mousse with the reserved banana and pineapple as described in the recipe.*

Tie a double band of baking parchment around a 2½-cup soufflé dish, to come 2in above the rim. Peel and chop one banana and place it in a food processor with the cottage cheese. Process until smooth.

Drain the pineapple, saving the juice, and setting aside a few pieces for decoration. Add the rest of the pineapple to the mixture in the processor and process for a few seconds until finely chopped.

Pour 4 tbsp of the reserved pineapple juice into a small heatproof bowl and sprinkle the gelatin on top. When spongy, place over simmering water, stirring until the gelatin has dissolved. Stir the gelatin quickly into the fruit mixture. Whisk the egg whites to soft peaks. Fold them into the mixture. Tip the mousse mixture into the prepared dish, smooth the surface, and chill until set. Carefully remove the paper collar. Slice the remaining banana and use it with the reserved pineapple to decorate the mousse.

# BOSTON BANOFFEE PIE

*Guaranteed to bring a grin to diners' faces, this is a winning combination of bananas and toffee.*

**Makes an 8in pie**

*8in cooked pastry case, cooled*

*2 small bananas, sliced*

*a little lemon juice*

*whipped cream and grated dark*
*   chocolate, to decorate*

**For the filling**

*½ cup butter*

*½ x 14oz can sweetened condensed*
*   milk*

*⅔ cup brown sugar*

*2 tbsp corn syrup*

### COOK'S TIP

*To make the pastry case, rub*
*½ cup butter into 1¼ cups all-*
*purpose flour. Stir in 4 tbsp*
*superfine sugar and press into*
*an 8in flan pan. Line the*
*dough case with crumpled foil*
*or baking beans. Bake at*
*325°F for 20–25 minutes.*

Make the filling. Place the butter, condensed milk, brown sugar, and corn syrup in a large non-stick saucepan. Heat gently, stirring occasionally, until the sugar has dissolved.

Bring to a gentle boil and cook for 7 minutes, stirring all the time (to prevent burning), until the mixture thickens and turns a light caramel color. Pour into the cooked pastry case and leave until cold.

Decorate with the bananas dipped in the lemon juice. Pipe a swirl of whipped cream in the center and sprinkle with the grated chocolate.

# TROPICAL BANANA FRUIT SALAD

*Not surprisingly, bananas go particularly well with other tropical fruits.*

**Serves 4–6**

1 medium pineapple

14oz can guava halves
   in syrup

1 large mango, peeled, pitted,
   and diced

2 medium bananas

⅔ cup preserved ginger, plus 2 tbsp
   of the syrup

4 tbsp thick coconut milk

2 tsp sugar

½ tsp freshly grated nutmeg

½ tsp ground cinnamon

strips of fresh coconut, to decorate

---

**COOK'S TIP**

*For an appealing decorative touch, use two small pineapples. Cut them in half, through the leaves, carefully scoop out the pulp, and use the shells as containers for the tropical fruit salad.*

Peel the pineapple, remove the core, cut the flesh into cubes, and place in a large serving bowl. Drain the guavas, reserving the syrup, and chop them into dice. Add the guavas and mango to the bowl. Slice one of the bananas, and add it to the bowl.

Chop the preserved ginger and add it to the pineapple mixture. Mix together lightly. Pour the ginger syrup into a blender or food processor. Add the reserved guava syrup, coconut milk, and sugar. Slice the remaining banana and add to the mixture. Blend to a smooth, creamy purée.

Pour the banana and coconut purée mixture over the fruit, adding the freshly grated nutmeg and ground cinnamon. Serve the fruit salad chilled, decorated with strips of coconut.

# BANANA HONEY YOGURT ICE

*Smooth and silky, this delicious banana ice is very refreshing when eaten after a rich meal.*

**Serves 4–6**

4 ripe bananas, roughly chopped

1 tbsp lemon juice

2 tbsp clear honey

generous 1 cup Greek-
   style yogurt

½ tsp ground cinnamon

crisp cookies, flaked hazelnuts, and
   banana slices, to serve

**COOK'S TIP**

*Switch the freezer to the coldest
setting about an hour before
making the yogurt ice to ensure
that it freezes quickly.*

Place the bananas in a food processor or blender with the lemon juice, honey, yogurt, and cinnamon. Process until smooth and creamy.

Pour the mixture into a suitable container for freezing and freeze until almost solid. Spoon back into the food processor and process the mixture again until smooth.

Return the yogurt ice to the freezer until firm. Before serving, allow the ice to soften at room temperature for 15 minutes. Scoop into individual bowls and serve with crisp cookies, flaked hazelnuts, and banana slices.

# BANANA AND PASSION FRUIT WHIP

*Creamy mashed bananas combine beautifully with passion fruit in this easy and quickly prepared dessert.*

**Serves 4**

2 ripe bananas

2 passion fruit

6 tbsp fromage frais

⅔ cup heavy cream

2 tsp clear honey

shortcake or ginger cookies, to serve

### COOK'S TIP

*Look out for cans of passion fruit (or grenadilla) pulp. Use with sliced banana and whipped cream to make a marvelous topping for pavlova or for sandwiching together individual meringues.*

Slice the bananas into a bowl, then, using a fork, mash them to a smooth purée. Cut the passion fruit in half. Using a teaspoon, scoop the pulp into the bowl. Add the fromage frais and mix gently. In a separate bowl, whip the cream with the honey until it forms soft peaks. Carefully fold the cream and honey mixture into the fruit. Spoon into four glass dishes and serve the whip at once, with the cookies.

# BREAD AND BANANA YOGURT ICE

*Serve this tempting yogurt ice with strawberries and cookies, for a luscious and light dessert.*

**Serves 6**

*2 cups fresh whole wheat
    breadcrumbs*

*⅓ cup brown sugar*

*1¾ cups ready-made cold custard*

*5oz fromage frais*

*⅔ cup Greek-style
    yogurt*

*4 bananas*

*juice of 1 lemon*

*¼ cup confectioner's sugar, sifted*

*½ cup raisins, chopped*

*pared lemon rind, to decorate*

*fresh strawberries, halved, to serve
    (optional)*

**COOK'S TIP**

*Transfer the ice to the
refrigerator about 30 minutes
before serving to allow it to
soften a little. This will make it
easier to scoop neatly so that it
looks attractive when served.*

Preheat the oven to 400°F. Mix the breadcrumbs and brown sugar in a bowl. Spread the mixture out on a non-stick cookie sheet. Bake for about 10 minutes until the crumbs are crisp, stirring occasionally. Set aside to cool.

Meanwhile, mix the custard, fromage frais, and yogurt in a bowl. Mash the bananas with the lemon juice and add to the custard mixture, mixing well. Fold in the confectioner's sugar.

Pour the mixture into a shallow, freezerproof container and freeze for about 3 hours or until mushy in consistency. Spoon into a chilled bowl and quickly mash with a fork to break down the ice crystals.

Add the breadcrumbs and raisins and mix well. Return the mixture to the container, cover, and freeze until firm. Serve with the strawberries, if using, decorated with lemon rind.

# BANANA FRUIT PUNCH

*This pleasingly quick and simple Caribbean specialty is perfect for a summer party.*

**Serves 3–4**

2 bananas

4 tbsp ginger syrup

½ tsp almond extract

½ tsp vanilla extract

4 cups mango juice

3¾ cups pineapple juice

1 cup lemonade

freshly grated nutmeg

lemon balm and orange slices, to
    decorate

Chop the bananas into ½in pieces. Place them in a blender or food processor. Add the ginger syrup and extracts and process until smooth.

Transfer the mixture to a large punch bowl. Stir in the mango and pineapple juices, then pour in the lemonade. Finish by sprinkling in some grated nutmeg. Serve chilled, decorated with lemon balm and orange slices.

**COOK'S TIP**
*Prepare this punch up to
2 hours in advance of serving
and chill until ready to
decorate and serve.*

# DARK RUM AND BANANA PUNCH

*The inspiration for this punch came from Guyana, where bananas, rum, and unrefined sugar are used to make a variety of potent drinks. Use any combination of fruits to decorate the punch.*

**Serves 4**

⅔ cup orange juice

⅔ cup pineapple juice

⅔ cup mango juice

1 cup dark rum

a shake of angostura bitters

freshly grated nutmeg

2 tbsp brown sugar

1 small banana

1 large orange, sliced

**COOK'S TIP**

*Dark rum from Guyana is traditional, but you can use white rum, if you prefer. For an alcohol-free version, use American dry ginger instead.*

Pour the orange, pineapple, and mango juices into a large punch bowl. Stir in ½ cup water and add the rum, angostura bitters, nutmeg, and sugar. Stir gently for a few minutes until the sugar has dissolved.

Slice the banana and stir gently into the punch. Float the orange slices on top. Chill and serve with ice.

# INDEX

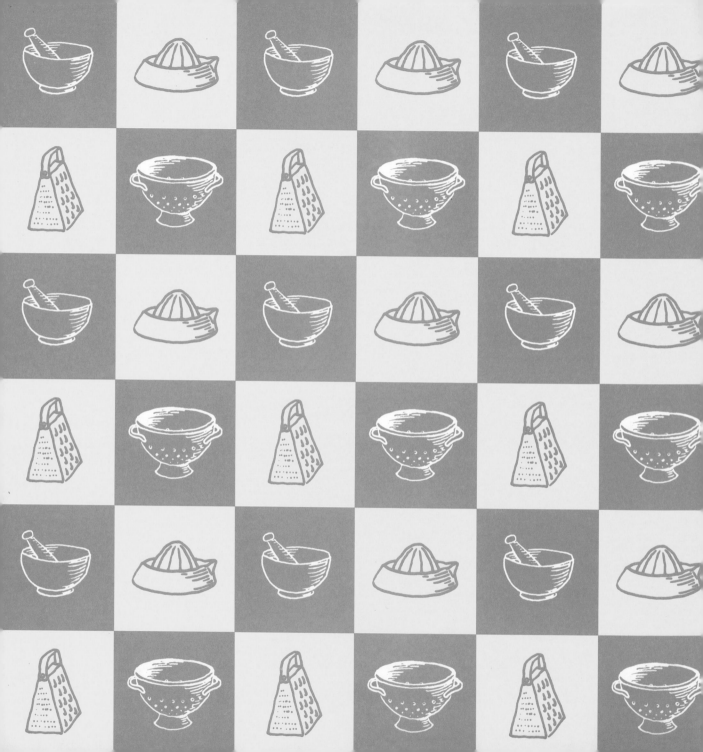